Today is a lovely day and
I've been improving it.

~ Florence Merriam Bailey

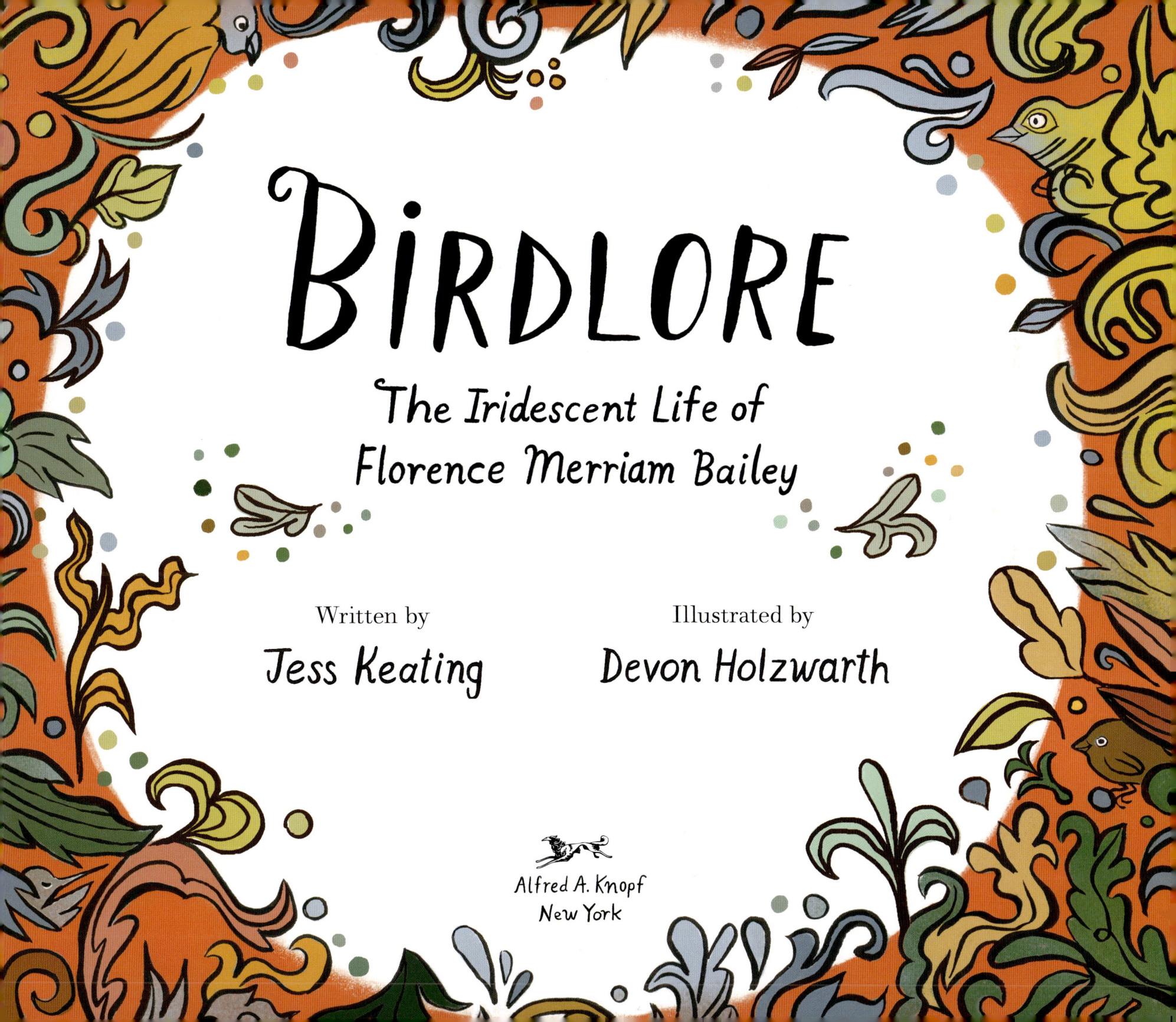

Birdlore

The Iridescent Life of Florence Merriam Bailey

Written by

Jess Keating

Illustrated by

Devon Holzwarth

Alfred A. Knopf
New York

Florence pulled up her thick wool stockings, sprinkled a handful of sunflower seeds into her coat pocket, and tucked Ruth under her arm.

It was time for adventure.

It was time for nature.

And best of all, it was time for *birds*.

She nestled into a copse
of locust trees.

With the violets, tiger lilies,
and trilliums at her feet,

Florence waited.
Florence watched.
Until finally . . .

They were here!

From season to season, Florence explored with her feathered companions.

She shared berries with bold and bright blue jays,

she strutted in the deep snow with ruffed grouse,

and she danced to the rhythm of woodpeckers drumming on tree trunks.

And then, under the veil of night and a bundle of warm blankets, she would explore the constellations with her mother while owls swooped through the trees.

If she had powerful wings like an owl, could she find her way to the stars?

To Florence, nature wasn't just something that surrounded her.
The mottled rocks, towering, twisting trees, and winding streams . . .
they were *home*.
The creatures that hopped, crawled, and swam sparked her curiosity.

But those with wings and a song? They won her heart.
Florence wanted to fill her life with the birds she loved.
So Florence found a way.

Tagging along after her big brother, Hart, she began to study her favorite birds. Florence didn't care that girls weren't usually scientists. If *he* could learn about birds, why couldn't *she*? Together, they collected specimens, took notes . . .

. . . and even learned how to preserve the animals for display.

Florence loved this time with her brother, but something wasn't right. Scientists like him learned about birds by examining their bodies, feathers, and eggs.

But Florence didn't want to examine dead birds in a collection. Birds belonged in the sky, flying free!

With a notebook in her hand and wonder in her heart, Florence set outside—this time to *learn*.

Florence discovered that her birds all had personalities of their very own.

Some were loud. Some were shy. Some were even silly! Her birds were a living rainbow of sights, sounds, and spirits, and Florence took careful notes on every detail.

Soon she began teaching classes in the renovated tavern that belonged to her grandfather.

She taught her friends about chickadees and warblers, robins and ruffed grouse, hermit thrushes and kingfishers.

Every new bird she met opened another door inside her heart.

When Florence was old enough, she left home to attend college. The city of Northampton didn't have as many trees and streams as the forests of Homewood, but it did have lectures on art, music, geology, chemistry, and literature.

Florence made friends with many other women at the university.
At supper one night, her classmate was excited to share what she'd
seen in the city.

"Thirteen birds—each with blue feathers, a white collar,
and a curious topnotch!"
But the beautiful birds weren't spotted on a tree limb in
the wild.
The woman had seen them *perched dead* on a hat.

Kingfishers! Florence thought. Kingfishers didn't belong on ladies' hats! They belonged at the water's edge, showing off their iridescent feathers as they hunted for fish.

Florence's heart thumped, and she bit back a cry.

FINE LINEN & FEATHER EMPORIUM

DRY GOODS & MILLINERY

SCHNITTER

Florence knew she must protect the birds from such horrors. But how? If only she could share her love of birds with the world. Then *nobody* would ever wear them as decoration on their silly hats again.

An idea began to hatch inside Florence's mind.

She cut articles from newspapers, shared pictures, and organized meetings.
She taught her classmates the beauty of bird *life*, and how it was so much more
stunning than any dead bird on a hat.

Her love of birds flew from one girl . . .

to another . . .

and another.

The more they learned about birds, the more Florence's classmates
wanted to help. Feathered hats were returned en masse,
and soon the campus was in a flap at the injustice.

Florence began to wonder. Sharing pictures of birds had changed so many minds and hearts, but was it enough? Wouldn't a *real* bird be even better than a picture?

She thought back to her brother's experiments and research. A dead bird in a lab was no better off than a dead bird on a hat.

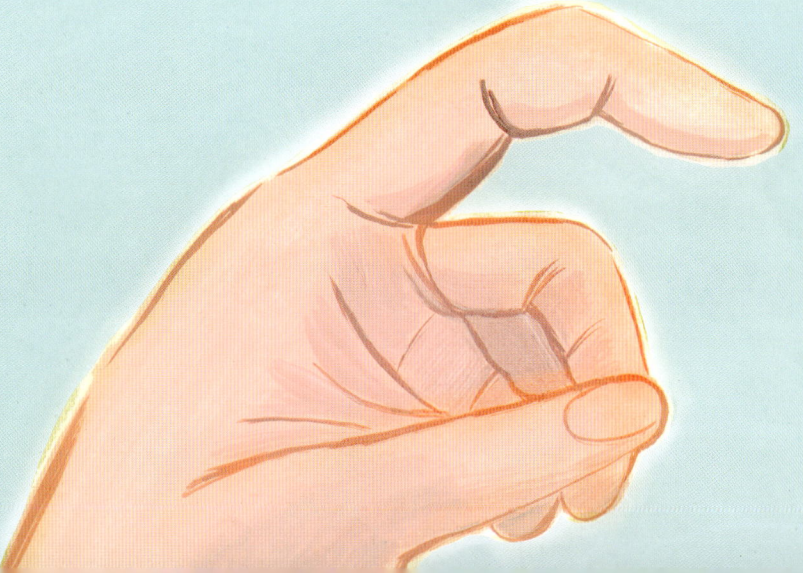

Florence wanted others to see birds for what they *really* were.

Vibrant. Colorful. Spirited.

And most importantly . . . *alive*.

A small flutter of wings began to beat in her heart, and her imagination took flight.

At half past five on a mid-May morning, Florence took her first group of women out into the forest. They each brought a notebook, a pencil, and some opera glasses to spot birds from afar.

It was time for adventure. It was time for nature.

And it was time to share her beloved birds where they truly belonged.

Florence watched.

Florence waited.

Until finally . . .

They were here!

With enchanted eyes, she and
the other ladies watched redstarts
and bluebirds,

cardinals and waxwings,
kestrels and kingfishers.
Flying. Singing. Preening.
Soaring.

Word migrated quickly. Soon Florence was leading
bird walks three days a week with students, scientists,
and professors. With every step, she learned more about
birds, and more about people, too. Through the eyes of others,
Florence experienced her birds in a different light, making
new discoveries about old friends. And seeing the birds
in real life changed people's hearts. They didn't just love
them—they wanted to protect them, too.

Each morning, Florence watched the skies with a joyful heart.

She'd shared her birds with the whole campus. But to protect *all* birds, she would need to share them with *everyone*.

Florence knew the way.

She pored over the notes she'd written through the years—
seeds of wisdom from countless bird walks.

Each page told the story
of a precious friend—

a steadfast star in the
constellation of her life.

If she couldn't bring
the world to her birds,
she would bring her birds
to the world.

Her field guide, she decided, would be the first of its kind. Page by page, she would take everyone on a bird walk of their own, helping them to learn, to lead, and to love.

And with the stirring song of a hermit thrush in her heart, Florence began to write.

Author's Note

Florence Merriam Bailey was born with birds in her soul. To her, it wasn't enough to love them—she knew she had to *share* that love with everyone she met.

At that fateful dinner party in 1886, Florence was upset to learn of the thirteen dead king-fishers that had been spotted on a hat, and the reality was much worse than she feared. At the time, "plume hunters" were killing five million birds every year for ladies' hats. One ornithologist even counted forty unique bird species on a stroll through Manhattan—all mounted on ladies' hats.

To me, Florence embodies what it means to love nature. We mustn't just enjoy the wild places and creatures that have our hearts. We must stand up, share what we know, and act to protect them. By taking others outside with opera glasses to see birds in their natural environment, Florence pioneered modern bird-watching. She is one of the most influential ornithologists of all time, yet her name is hardly known.

It wasn't long after her popular "bird walks" in college that Florence decided to write her first book. Birds weren't just for scientists, students, or museums. Birds were for *everyone,* and therefore Florence wanted to write something for everyone, including women, families, and children, to enjoy.

Birds Through an Opera Glass was published in 1889, and it's known as the first field guide to American birds. Unlike most women of the time, Florence chose to publish under her real name—standing up for both birds and women. Her work even led to the passing of the Lacey Act in 1900, which helped protect animals from illegal trade.

Florence traveled for more than forty years with her husband, Vernon Bailey—sharing her love and knowledge of birds at every turn. She led nature walks and gave bird talks wherever she went and wrote several more books in her lifetime.

In 1948, after eighty-five years of bird-watching, bird writing, and bird saving, Florence's soul took wing. And thanks to her life devoted to the birds, Florence found her way to the stars.

Become a Bird-Watcher with Florence

Birds live on every continent and are some of the most beautiful and diverse creatures on the planet. You don't need to be a scientist to study or enjoy birds—you can start today right in your own backyard, in the schoolyard, or in your favorite park.

- Bring a notebook, a pencil, and some colored pencils or markers. Bird-watchers often keep notes to remember all the beautiful birds they spot, but you can also draw or sketch them! Birds make excellent subjects for artists. These firsthand notes and drawings are called *observations*.

- Start by walking as silently as you can. Birds can be scared by quick movements, so speaking only in a low, quiet voice can help. Every so often, stop walking and let yourself be perfectly still. Do you notice any bird friends yet?

- If you can't see any birds, find a quiet spot by a tree and close your eyes for a moment. Often, you can hear birds before you see them! Listen carefully. Do you hear any chirps, whistles, squawks, or caws? There are many websites that can help you identify bird sounds. Visit allaboutbirds.org to explore more.

- When you do spot a bird, be very still. Can you notice a few details about it? What color is it? How big is it? What unique markings do you see? Make some notes about the bird in your journal. Later, you can use a bird guide from the library to help you identify it.

- If you really love birds, like Florence, you can even make your own field guide. Assemble your notes and drawings, and see if you can include one or two facts about each. Be sure to share your field guide with someone who loves birds, too!

Florence's Feathered Friends

Indigo Bunting

Rose-Breasted Grosbeak

Belted Kingfisher

Blue Jay

American Kestrel

Cardinal

American Robin

Red-headed Woodpecker

Northern Flicker

Cedar Waxwing

Bluebird

Crow

Chickadee

Bank Swallow

Ruby-throated Hummingbird

Common Grackle

Tufted Titmouse

Red-eyed Vireo

Catbird

Chipping Sparrow

Scarlet Tanager

Red-winged Blackbird

American Goldfinch

Redstart

Pileated Woodpecker

House Sparrow

White-breasted Nuthatch

Barn Swallow

Ruffed Grouse

Kingbird

Purple Finch

Great Horned Owl

Hermit Thrush

Barn Owl

Yellow Warbler

Florence's favorite bird species can be found flying, foraging, and nesting throughout the book.
Can you spot them all?

To Steph FC and Justin—
my favorite bird people
—J.K.

For the crows of Rabental
and the blackbirds of the Hollandwiese
—D.H.

THIS IS A BORZOI BOOK PUBLISHED BY ALFRED A. KNOPF

Text copyright © 2025 by Jess Keating
Jacket art and interior illustrations copyright © 2025 by Devon Holzwarth

All rights reserved. Published in the United States by Alfred A. Knopf,
an imprint of Random House Children's Books, a division of Penguin Random House LLC, New York.

Knopf, Borzoi Books, and the colophon are registered trademarks of Penguin Random House LLC.

Visit us on the Web! rhcbooks.com

Educators and librarians, for a variety of teaching tools, visit us at RHTeachersLibrarians.com

Library of Congress Cataloging-in-Publication Data is available upon request.
ISBN 978-0-593-48858-4 (trade) — ISBN 978-0-593-48859-1 (lib. bdg.) — ISBN 978-0-593-48860-7 (ebook)

The text of this book is set in 15.5-point Walbaum MT Pro.
The illustrations were created using a combination of gouache and watercolor with color pencil and Procreate.

Editor: Katherine Harrison
Designer: Nicole de las Heras
Copy Editor: Artie Bennett
Managing Editor: Jake Eldred
Production Manager: Claribel Vasquez

MANUFACTURED IN CHINA
10 9 8 7 6 5 4 3 2 1
First Edition